PRAISE FOR

"How could poetry about violent assault, about homophobia, about dysmorphia and more all sparkle so brightly? Ari Lohr's innovative language – and his charting of it – surprises reader skin like faraway fireworks. At just nineteen years old, he deals with heaviness with such light ease – and in *Gravity*, in such tight spaces, he's made us a banging breathtaker."

– Pascale Potvin, Editor-in-Chief of *Wrongdoing Magazine*

"In Ari Lohr's stunning poetry collection *Gravity*, he explores a myriad of themes with vulnerability, urgency, and care, from the unspeakable violence committed against queer boys, to an ode to xyr teeth, to the tender ache of love poems penned to their lover Adam. I was left stunned and amazed by how he revealed and unspooled xyr words on the page, exploring the limits of single-word lines and sporadic, scattered phrases like staggered thoughts against the white. I'm eager for all readers to hold this collection in their hands soon, and for Ari to continue writing what they do best: telling stories through poetry and fearlessly making them known."

– Sofia Aguilar, poet and author of *STREAMING SERVICE: golden shovels made for tv* and *STREAMING SERVICE: season two*

"*Gravity* holds unyielding forces in its words: the crushing weight of an ever-expanding, ruthless universe, and death and love with the equal pull of their inevitability. Lohr navigates it all through a speaker who 'is always reaching back' while falling through these principles. Just like the collection named after it, Lohr confesses: 'gravity is a silent killer.'"

– Sunny Vuong, Editor-in-Chief of *Interstellar Literary Review*

Gutslut Press

ALSO BY ARI LOHR:

EJAY.

GRAVITY

Ari Lohr

Gravity
Copyright © Gutslut Press, 2022

All rights reserved. This book may not be reproduced or used for any purpose without express written permission of the author.

Cover art by Ari Lohr

first edition

ISBN: 978-1-387-61189-8

gutslutpress.com

light birthing heat
heat birthing light
wake me
god
wake me
from this impossible dream

(four facts about space)

i.

Each year, gravity pulls the moon 1.6 inches further from Earth. You won't live to see it, but the moon will eventually abandon us, leading to changes in the earth's tilt, falling sea levels, extreme temperatures, and mass extinction. Everything is quiet in space; gravity is a silent killer.

ii.

According to the law of conservation, matter can neither be created nor destroyed. Applying this law to consciousness gives us the following conjecture: you don't disappear; you don't die. You just change form.

iii.

No one knows how the universe began; no one knows how it will end, if it will end. Compelling theories exist, but scientists don't have a definite answer.

iv.

Outer space is an ever-expanding, empty void, with only 0.0000000000000000000042 percent of it containing any actual matter.

Let go, if you can, of your worries – most of this is nothing.

Everything that is, everything that ever was, everything we once knew, begins and ends in darkness. Can you fathom that?

TABLE OF CONTENTS

(four facts about space)	vii
TABLE OF CONTENTS	ix

i.
LOVE

THE LAW OF CONSERVATION	3
part one	
FALLING	5
DRIVING HOME FROM A HIKING TRIP	9
BEFORE YOU LOVED ME, I HATED MY TEETH	12
GETTING HIGH IN YOUR CAR	18
I COULD	21

ii.
RAPE

THE LAW OF CONSERVATION	25
part two	
A SELF-ERASING ELEGY TO MY RAPE	27
in four parts	
FROM THE INTERNET // FROM MY NOTES APP	28
ERASING MY RAPE	29
in four parts	
10PM ON DARTMOUTH AVE	30

ERASING MY RAPE	31
NO	32
	33

iii.
DEATH

THE LAW OF CONSERVATION	35
part three	
A SONG FOR THE GAY BOYS WHO NEVER CAME HOME FROM PRIDE	37
part one	
YOU'RE LEAVING FOR CALIFORNIA, AND	44
THEY HID IN A CLOSET	46
BEFORE YOU LOVED ME	49
A SONG FOR THE GAY BOYS WHO NEVER CAME HOME FROM PRIDE	52
part two	
(interlude)	lv
A SONG FOR THE GAY BOYS WHO NEVER CAME HOME FROM PRIDE	56
part three	
UNFOLD	58
CONTRAPUNTAL TO NOTHING	62

(ABOUT SPACE)	lxiii
GRAVITY	64

iv.
NOTHING

THE LAW OF CONSERVATION	71
part four	
ESSAY ON LEAVING	73
after Hunter S. Thompson	
NOTHING AND THEN NOTHING	80
ELEGY IN FOUR ACTS	82
NOTES	87
ACKNOWLEDGEMENTS	89

i am never more dangerous than inside

the arms of a man

who will die

before me

— Sam Sax, *Politics of Elegy*

i.
LOVE

She drags the deer across the forest floor. Mom hunts and I follow. A cruel kind of inertia.

A bullet hole, a third eye, on its stomach. Cold metal blending with flesh, the deer's life melting away, sharing its body heat with the bullet, the air, Mom's calloused hands. This is how it goes, Mom says, when I ask her about death. This is all I know.

The clouds part; Mom sweats beneath the sun. Light birthing heat.

The deer shivers beneath her grip.

FALLING

Kissing me in public without

      looking over your shoulder first.

            Dipping me in your

      wedding ring arms on comm ave.

Telling you I love you the

      first time outside a poem, the

            first time outside.

      Once, I saw a bottomless pit and

jumped, which is to say

      that I am always falling for you, that

            I am always falling.

      We get high in your car and

I babble about some random dream I once

      had, a slurry half-thought-out conjecture about

            religion, how

I imagine the final circle of hell to

      be nothing, nothing and

then nothing, nothing but

a bottomless pit in

which you keep falling, nothing but

an empty void, a

cruel alchemy, kissing your

boyfriend in public and

pretending you're both alone, ignoring the

man glaring at you across the street,

writing a poem about it later, queer

love in the space of eleven lines, queer

love in a starbucks parking lot, queer

love in the lonely language of ink, queer

love in the home of the

heart's most violent vocabulary, queer

love in a thousand different dialects, queer

love as a bottomless pit, queer

love as the act of falling, queer

love as hell's holiest circle. god becoming

flesh. Boy becoming

    blasphemy. Entering my

        empty void, our tongues tangling like

    two passionate snakes. Take me to a place where

everything burns. Take me to a place beyond

    where bodies can go. Take me, god, take me

        anywhere. Love me like the

world is ending, until the

world ends, like the

    world will never end.

        Kiss me

until your body decomposes

beside mine, clouds of

    smoke whiskering the

        night sky,

    ghosts of

queer

        boys

                haunting

        the

city

        beneath

                us.

DRIVING HOME FROM A HIKING TRIP

The car stops
on the edge
of I-5.

Mom's knuckles
turn white. The wheel shrivels
beneath her grip.

Dad yells
just loud enough
to drown out her sobs.

In the backseat,
I notice something
peculiar

while looking out
the window
to distract myself:

From this far away,
the Cascades look like one thing
and one thing only –

An upside
down egg carton.

I confirm this
by counting the rise
and divot of each mountain
with my fingers.

Mt. Rainier: Egg.
Mt. Saint Helens: Egg.
Mt. Hood: Egg.

I know it's dumb,
but from that moment —
from the moment
that second grader
looked out his window
and saw nothing but eggs,
swore he saw

nothing.

but.

eggs.

I imagine
some god cracked a laugh
at my pain
and how I only remember
its dumbest,
most innocent parts:

The sky
and its open mouth.
The Columbia River
running like an ever-
broken yolk.
The driver
honking at us,
his trailer carrying
hundreds of chickens
in tiny cages.

And just then –
in the middle
of motionless hell,
Dad sees my fingers
touching the glass,
stops yelling at Mom
to yell at me,
and Mom sighs
and starts driving
again,
leaving
whatever they were fighting about
to fry
on the road
behind them.

BEFORE YOU LOVED ME, I HATED MY TEETH

I was born
with a lip incompetence,
meaning
– at its relaxed state –
my mouth
rests slightly ajar,
my upper lip
raised above
my buck teeth
in the shape of
a triangle.

Growing up,
the neighborhood kids
couldn't decide
between calling me
Dorito
or *Squirrel*.

When you have
body dysmorphic disorder,
you cope
with each trigger
with comedy –

like
when people ask
how it is
to have dysmorphia
I tell them
I quit my job at subway
because I was irrationally angry

at the chip rack,
that when I say
I have nuts in my mouth
you can't tell
if I'm giving a blowjob
or simply saw my face
in a mirror.

Between laughter,
I do not speak
of how
I purse my lips
when I open
my front camera.

How I bought
an N-95
three years
before the pandemic.

How I moved to Boston
so I could justify
covering my mouth with a wind mask.

When you ask
why I joke
about my face
so much,
I want to say
dysmorphia
is a language
of unspoken shame,
that since my teeth
do not
fit

in my mouth
my tongue
bends
each bone
to be beautiful
that my jaw
is only pretty
in a punchline

except
when I do speak
nothing comes out
except

Hey Adam,
what do you call
an anorexic
with buck teeth?

A rake.

When you figure out
what the joke means
I've already found
three excuses
to end the conversation,
because it is easier
than explaining

why I never smile
in family photos

or why I wear
a mask in my own home

or why I begged my mom
to get braces

or why
I fucking *hate*
Alvin-and-the-Stupid-Fucking-Chipmunks.

I am so good
at biting my tongue
but so bad
at keeping my mouth shut.

I know I'm freaking out
over nothing –
but when I'm eating a cupcake
and see
my teeth marks in the frosting
I choke
on my own
disgust.

When I walk past a mirror
I stop
and count each atom
between my lips
until I've swallowed
my whole throat
and pulled
each tooth
like a thread.

When I say
I never smile
in pictures
I mean

my mouth
is an ever-agape aperture
of flesh
my front teeth
are two flashbulbs
of buck-white enamel
I can't turn off.

Sometimes,
I leave the room to eat
so you don't see me
chewing.

How do I tell you that
and not sound insane,
Adam?

How do I tell you
that the first time
I knew I loved you
was on our third date
when you told me
that this —
this
was your favorite version of me:

When my guard is down.

When my jaw is dropped.

When my mouth hangs open.

You say
you know
that something

is always beautiful
when we're together,
because
I'm always gasping
in awe.

You say
I'm basically just Sandy Cheeks.
Only, cuter.
And a twink.

You say
you love it
when our teeth
clank together
while kissing,
that you only loved me more
when I stopped
pausing
to apologize.

And now,
I smile with my teeth.
I take my mask off.
I eat doritos.

Something
is always beautiful.

I know I shouldn't do this.
I know I shouldn't love myself
just because you love me.
But god damnit Adam,
god damnit,
I do.

GETTING HIGH IN YOUR CAR

You sit in the driver's seat
and watch as I clumsily
hold a joint to my lips.
I cough eight times,
coating the windows
with frantic
puffs of smoke.
The universe is always expanding.

The universe is always expanding and
you are motionless
as I rave about vegan chorizo and
personal-sized pizzas
and you graze my cheek and
half-chuckle about
how my favorite restaurant
is called *blaze pizza*,
how you're *not really*
grazing my cheek,
how electrons repel each other
and make it impossible
for two bodies to touch,
how with each passing second
– no matter how still we sit –
the universe expands and
we drift hundreds of atoms apart.

We could be in this car for hours,
going nowhere like
we were, and
have lived
thousands of different
lives, each one trapped

in endless and
aching motion, too
high to remember any
of it.

I don't smoke or drink or write
to forget,
but I pity my memory.
I pity
how my lips trace strangers' outlines
more often than they learn their names,
how I sit in the same car
and share the same air
as my lover
and imagine myself
writing the same poem about death
and the universe and the space between here
and the inevitable,
how I see the same
dizzying void
as I imagine our future too.

The little house by comm ave.
The pet fox and the flower garden.
The lives we take for granted
as smoke fills the car
and our bodies inch
impossibly close to
each other.

This
is death
before death.
This is loving you to the moon

and then some,
the rituals and doctrines
we uncover and keep close,
knowing
that the whole universe
could be high
on this love
and I would never stop
loving you.

This is muscle memory,
the symphony
of darkness
and calloused hands.
The universe is always expanding

and I am always
reaching back.

I COULD

have eaten them —
the bagels, I mean.

They went stale yesterday
but
I could have eaten them.

Today — mold.

I could have eaten them

except

they come in packs of six and for some reason have a shelf life
of four days meaning i would have to eat one and a half bagels
a day which is a dilemma i did not consider when i bought this
pack of bagels last week
and of course

I will *not* half-ass a bagel.
So
instead
I leave them blue
and pungent
and spotted
like
a northern mockingbird's eggs.

•

I could

have seen you before you died,
could have
bought a plane ticket
or called
or sent a letter
but
I was at work or school or
busy studying –
busy.

speak this unspeakable truth.
speak this cruel, mundane language.
speak
till you are stale
and restless
and nothing
but compost.
in enough time,
all decay looks the same –
blue
and pungent
and so,
so

                        unfinished

ii.
RAPE

Before dinner, Dad strips the meat from the deer's hide, slapping it on the grill. Mom finds comfort in this meditation: the calculated exchange of heat, the glowing coals, the trembling flame, being becoming bread, playing god, Dad's divine and magic hands.

Venison, green beans, and scalloped potatoes. Dad shuffles a bite like a pinball between his cheeks, blowing air to cool it down. His face turns red. Soon, his body will digest the meat, converting it to heat, movement, and fat.

He swallows and takes another bite.

A SELF-ERASING ELEGY TO MY RAPE
in four parts

i.

Everything. I wrote down everything. You raped me and I wrote down everything. I'm not lying.

There's nothing left to say. Nothing at all. Nothing in my notes app, nothing in my journals. Nothing about that man, that place, that memory, nothing I haven't written. But (of course) that means nothing. The last time I wrote about it was a reddit post. I wrote down everything and asked if it "counted" as rape. All I got was republican men gaslighting me and poets saying it was too "on the nose."

Yet here I am again, writing that place to life, keeping the thread undone. I exploit my memory, my trauma, constantly: the hardwood floors, the white bedsheet, the cigarette butts in the corner, the aroma of smoke. You persist in a cocktail of ink-infused rage.

I hate what I write. All of it. It is nothing but shit.

Dear god, let me leave this place, leave this memory behind. Dear god, let me leave. Let me leave. Let me forget the truth. Let me forget this all. Let this all be nothing. Nothing, god, that's it. Let this all be nothing, nothing, nothing but endless nothing.

i am starting to   just bad sex   Forget details

when it happened
and stuff but like idk

like

i don't know

if it counts

I have social anxiety   I was hypersexual
embarrassed   to leave   to ask   to leave
he never asked

never asked   but   I   didn'
no

i knew   if i was   frozen
on top
disassociated   i was in "control" of the situation

he moved my hand   despite me kinda but sorta maybe not really idk
protesting

(???????)

I protested about his nails   sharp
his fingernails were sharp

my   breath still tastes   like   beer   my   breath still tastes   l
beer from his mouth

that would be a good poem topic tbh

ERASING   MY RAPE
in four parts

ii.

                                    You raped me
               . I'm not lying.

There's nothing left   . Nothing   in
                                            that
place, that memory, nothing   But
          nothing.

      I am
  undone   my memory
                                                   in
              smoke. You persist
   .

I hate what I write.   It is nothing but shit.

Dear god, let me leave this place, leave this memory
       ,   leave.
               Let this all be nothing   Nothing, god,
                          nothing but endless
nothing.

the male appendage / the drunk libido / your calloused hands / my chapped lips / round / like a quarter / like a clock / like a bullet hole / like a wedding ring / you said / *i'll give you 40 to finish in your mouth / look at those big-ass teeth / how about 50 if you don't bite* / i said nothing / when you entered me / nothing except / *fingernails fingernails fingernails* / sharp / like shattered glass / like barbed wire / like a silent vow / like a lost language / *god fuck i wish this was Adam* / i said nothing / *god fuck i wish i was dead too* / when you entered me / i started to bleed / i said nothing / or maybe / maybe i said his name / maybe i said *Adam Adam Adam* / maybe you said *who's that?* / maybe i said / *no one / an ex / nothing / everything* / as you entered me / as i gasped in awe / as i started to bleed / as i said nothing / as you said you would finish in my mouth / as you finished in my mouth / as you gave me $50 / as i bought condoms / & lube / & a wind mask / later we will kiss / our teeth will clank together / i will say *i'm sorry* / i will write a poem / i will say nothing / i will bite my tongue / i will say / *leave me with this burden, give me your severed hands, take me to a place where everything burns*

ERASING MY RAPE

iii.

There's nothing in
　　　　　　　　　　　　　　　　　that
place,　　　　nothing　　　　But

undone　　　　memory

　　　　　　　．

what I write　　　is nothing　　　．

god, let me leave　　　leave
　　　　leave.
　　　　　　Let this all be
　　　　　　　　　　　　　endless

ologize if this isn't the right place to ask this, and **I'm sorry** for such a long post -- I'm
 sure where to turn or who to ask about this (I'm really disassociated right **no**w)
d use a stranger's perspective on it: Last night, I met someone on grindr -- I was look
 sugar daddy (**no**t too seriously haha, I'm a broke college student and was kind of tee
ersexual at the time [BPD symptom]) and this one person hit me up. We talked for a
I assessed the situation and it seemed relatively ok, so I decided to head over to
tment. The plan was to hang out and chat for about an hour, fool around physically,
t 50 bucks (tbh **I didn't** realistically expect to get any money, I'm **no**t that dumb, b
't really care, I was hypersexual and disassociated I think). When we were walking up
 to his apartment, I sort of popped back into reality mentally (for all my BPD peeps
, I suddenly got sexually repulsed) and was like wait, am I really doing this? but I k
ing up the stairs anyway. But at this point we were fooling around and it was whateve
the one to take off my belt as well as unbutton his shirt (before getting into bed). At
t though I was already naked and realized that I probably couldn't leave easily -- I h
d anxiety and I was too embarrassed to spend the 5 minutes getting dressed and grabb
tuff and asking to leave, etc. I don't think he would've harmed me or threatened me
d to leave, and that probably wasn't explicitly something that stopped me from asking
d leave, but that fear was probably in the back of my mind. Idk. But at one point
ed fingering me -- his nails were sharp and I said something along the lines of "ouch, y
 are too sharp" and he said "wrong angle" and kept going. At one point, he pulled
loms and lube from his nightstand drawer. I told him "I don't think I want to be fu
ght" and he huffed a bit and said "hmm, can I play with it at least?" I paused a bit
ly said "uhh.... a little." He started fingering me and I absolutely froze. I closed my ey
disassociated and I was completely soft the entire time. At a**no**ther point, I started giv
oral -- I was laying down, resting my head on his pillow, and he was on his knees with
 in my mouth. I had one hand on the base of his cock, and he moved my hand away
ed facefucking me. I didn't want to be facefucked, and he didn't ask if I did. I sor
ed my hand back to his cock and then he lightly pushed it away, and I leaned back a
ng to get his cock less deep in my mouth, I felt like I was about to gag) and he said
e on, you can take that." I awkwardly mumbled "haha I have to warm up first" (I had
 what to say lol) and I put my hand back on his cock to stop him from facefucking m
 remember well but idk if he moved my hand away again. Probably **no**t. I don't think
ut I k**no**w that he did facefuck me again once or twice later on. He never asked if it was
cefuck me, but I had initiated giving oral in the first place and didn't explicitly **say "no**
ucking, when he started to facefuck me. There were points throughout the night whe
ically moved on top of him (i.e. to start sucking / playing with his nipples, giving him
 a diff position, kissing his neck, grinding on him, etc). At the very beginning of
unter it was just because it was the one thing I found kind of hot (and it was the only
ld get semi-hard), but by the end I was only doing it because A.) I wanted to keep h
 fixating on the fact that I was soft (I was very embarrassed) B.) it was the only posi

                              iv.

  nothing
                                              But

  what I write   is

                                         endless

iii.
DEATH

Once again, I'm up late combing through research articles:

Up to thirty stars explode each second somewhere in the universe, releasing enough energy to illuminate the galaxy for weeks.

Heat birthing light.

A star's life ends when it consumes the entirety of its fuel and is no longer capable of burning.

This is how it goes: A third eye. An exchange of heat. Death. Inertia. A black hole. Everything.

Crushed by sulfur and iron, the star becomes so dense it collapses beneath the weight of its own gravity.

A SONG FOR THE GAY BOYS WHO NEVER CAME HOME FROM PRIDE
part one

    you shot him

when he was walking home
from pride,
just hours after
you sang
for your sunday church choir.

    tell me:

do you know
what it sounds like
to die?

have you heard
the song
of a last breath,
felt the symphony falling
from your lips?

today, i'm walking home
from my first pride.
i'm convincing strangers that
my boyfriend is my brother as
i write this poem
on the back of
an upside-down rainbow flag
and

i can't help but think
about the air

crescendo-ing in my lungs

how i am so afraid
of dying
i hold   my breath
so i never have to
take my last   one

how the pulse victims
thought the
gunshots
were

beats   to a   song

at first

how last year
a gay boy
from my hometown
was lynched
behind a church

                 his

                      legs

                               swinging
                      like
               a
                      metronome

how anti-gay organizations
       write songs
             to get more followers

how the god-given gunshot
of a church choir
pierces   each closeted gay boy's   ears
like nails   on a cross

how you *sang* before you shot him

how you baptized him
in his own blood
for an absent god
who doesn't even know
                     his name.

     tell me:

do you hear it too?

the sound of him
taking his last breath?

the sound of yours
as you grip   take aim   and pray?

     tell me:

have you heard
the sound of a dead boy
singing
beneath the dirt?
the orchestra
of his outstretched hands
praying
for the conductor
to finish the song
already?

       tell me:

do you know
the melody
of a dying
child?

do you know
the octave
of a mother's
scream?

do you know
what it's like
to form
    your
       lips
           into the shape
               of another
           dead gay boy's
     name?

to collect
dead queers
in your poems
like records?

to let them spill
from your mouth
like blood?

to paint
a target
on your back

and
        dance   to

                                  the   rhythm
              of

        bullets
                      flying
                              through
       the

           nightclub
                        air?

    tell me:

do you know what it's like
to make a ballad
from a bullet wound?

to march in parades
and harmonize
with the ghost
of every queer boy
who never came home
from pride?

to hear a   beat
and not know
if it's a bullet   a song   or your
own
pulse?

           do you know why we love music so much?

                    the beat reminds us we are still alive.

he was wearing headphones,
you know.

his music was so loud
he didn't even hear the gunshot.

the boy was 16
and the 3rd chair in his high school band.
he died in his mother's arms
as she held him
like a note
and prayed.

    tell me:

do you know
what it sounds like
to die?

it sounds like a prayer.

                like a song.

                              it sounds

                 like

                . . .

YOU'RE LEAVING FOR CALIFORNIA, AND

yeah, yeah, Adam, I know. I have daddy issues. I cried when you last fucked me. Nothing is permanent, everything is connected, nothing, everything, nothing, some other stupid esoteric non-dual spiritual shit. Christ, is this *it*? I'm begging you to stay and you're fiddling with a loose thread on your shirt. *please please please Adam i love you please* pull pull p u l l p u l l p   u   l   l   My god, are you *pulling* it? Have you *ever* worn clothes before? Fuck, Adam. Remember the night we first kissed? How we told each other everything, undressed each hope and fear and memory beneath the moon's warm glow? Ok, yeah. I know that was cliché. But I'm 16. I know this is pointless. But I love you. I know it's late, I know you're tired, I know your bus is coming soon or something, but *I love you*. I love you, Adam, my god, my god. Adam, my god. God, Adam, can't you see I'm trying? Listen to me. Your favorite food is bagels. You like science shit. You listen to pop music. You hate the beach. *Can't you see?* I know everything about you, everything. I know you. I know this isn't you. I know this, Adam, I know this, I *know* how this ends, I know, *I know*: the wind holding you like a breath, like a mother, gravity's invisible arms, your lungs swelling like balloons, your body baptized in blood, *endless and aching motion*, me at 16, 17, cobbling together poems from stupid, random, mundane shit, trying to make sense of this, trying to write a book, trying to let go, trying trying trying to keep the thread wound, to keep the memory alive, to keep you alive, Adam, to keep you alive, to keep you, to keep you. I'm scared, Adam. I'm scared of you dying. I'm so scared of you dying I'll kill myself before you can, I swear. Can't you see I'm desperate? Can't you see I'm trying? Can't you see? Can't you? I know this is pointless. I know I can't stop you. I know you don't believe me. I know you don't believe, in god or anything else, but fuck, Adam, fuck, this is what it means to pray, to be holy, to be everything, to be nothing, to be, Adam, to be, Adam, to *be*.

THEY HID IN A CLOSET

stitched together with transcripts of 911 calls from the pulse nightclub shooting

```
tell the officers         they found      a gun    by his rib
tell       the ambulance  i'm leaving my friends   behind
tell              the shooter    he already knows we're here
       tell me            god    do you know where you are
       be honest   say it again  it's not that hard to kill people
       it's not that hard to     push on your wound
                                       let him die this way
                            in                 the stomach
                                in                the chest
                                     in           the knee
                                     the operator said
                                              i can't imagine
                                              what you're
                                              going through
```

•

                imagine
          if you can   spraying bullets
      a ringing phone   a 911 call
an automated voice message   *ma'am, what was that banging?*
they are knocking on the door   no way out   without
            running   without running
                into him

•

  tell me   god
    can you see
     yourself
      in the background
       distorted backstage
        can you see
         he's coming
        can you hear
        someone yelling
      your name
     someone yelling
     my leg
  my leg
my leg
how fucked up is that
   god
    tell me what's happening
     tell me what it looks like
      above
      tell me
      where are you
       hiding
        this time
       did you notice
      everybody
     is hiding
      in the closet
       did you notice
        everybody
       is in the closet
        they are still there

BEFORE YOU LOVED ME

Mom would take me
to mcdonalds every saturday
while Dad and my brother
went hiking with their friend Craig.
I would order the usual

(six chicken nuggets, honey mustard, bbq sauce)

eat the skin off each one
then count the number of bites
based on how many nuggets I had left.

six nuggets left – i'll eat this one in six bites. five nuggets – five bites.
so on.

Like all things,
this ritual ended
when Craig moved to San Diego
and I grew older,
and my brother was busy
with school
& Dad was busy
too.

Sometimes,
I imagine myself –
adulthood just out of reach –
dissecting each nugget as if
the restaurant were a morgue,
and think there must be a metaphor in there
somewhere, that this nostalgia somehow means
something, that I could reconstruct this memory
and publish it in *Epoch* or *The New Yorker*

or someplace.

Yet another empty threat.

I know
none of this
was how I remember it –
Dad was an alcoholic
and Craig was a coke addict
and my brother had cancer
and Mom was always stressed with work
and the nuggets were always
slightly too chewy
for my teeth
which I would
hide
beneath
my shirt collar
between
each
bite.

do it.

write
your revisionist romance.

open your mouth
like a wound.

chew
on this memory
and make it

nothing.

swallow it
with your pride.

A SONG FOR THE GAY BOYS WHO NEVER CAME HOME FROM PRIDE
part two

. . .

or maybe
    it's
        a lingering breeze.
            A sunset.
                A sundown town.
                    The heat
                    falling from the boy's hands,
            a lone thread
                unwinding
            in
            the
              fabric
                of
            night.
Maybe
    it's
        the ashes.
        The flame.
        The stench
            of melting plastic.

The night after pride,
    three men sing church hymns
    while burning a rainbow flag.
    Some sick harmony
    of confetti   fear   and smoke.

A boy wearing headphones walks by.

I'm walking home from my second pride.

Across the street   a pair of clenched fists
                                 and a wandering gaze.

I let go of my boyfriend's hand.

●

I don't want to write of closets   but once
                                    I saw a boy
                                              turn a rainbow flag
                                                        upside-
                                                        down
                                                  so it didn't stick out
                                    of his bag.
                        Once
                                  I saw a man
                                              swear
                                      his boyfriend
                                      was his brother
              until each bus passenger's
shotgun
shell
       stare
                had vanished.
                        I am not here
                             to write the blood
                                    from a man's hands.
                                            This
                          is not a metaphor
                    for murder.
                This
     is

an elegy to noise / an unspeakable song / an undying light /

              this
     is
              what it
                       sounds like

to die / to pray / to spill from each other's wounds / to sing
beneath the dirt

(interlude)

Did you know
hate crimes are
highest in pride month?

According to experts,
gay men are more visible   at parades,
and therefore
more likely to be targeted.

                    IN OTHER WORDS

an obvious truth:

a rainbow flag
      won't stop
           a bullet.

                    IN OTHER WORDS

    the difference
        between a pride parade
             and a funeral
      is simply

           the genre of music.

A SONG FOR THE GAY BOYS WHO NEVER CAME HOME FROM PRIDE
part three

    What
    is the difference between

a pride parade   a concert   and a funeral?

    Last year,
    a gay boy from my hometown
    was shot.   Another was lynched.

The day before,
    the town had its first pride parade.
The boy held
         a plastic flag
             and marched
                 to the sound
                     of a snare drum

(his   heart   never   sk ipp ed a   beat)

        each step
             a lone thread
                     unwinding,
               the night
               silent
               as a   shotgun,
                     Florida
                  heat
                   falling
                    from
                   his
                     hands

as his body
went cold.

The boy died

listening to music   because of music   in spite of
music.

      Tell me:

Do you hear it too?

UNFOLD

It's all over
when

the world unfolds beneath us.

Wake me when
I reach the moon.

Wake me when
we become nothing.

We pretend
we are alone,
forgetting

we are strung along
by the same unwinding thread.

You hold me
as you tell me this.

I don't believe you
until
you leave.

You'll reach the moon
if you fold a piece of paper in half enough times.

I fold and keep folding.
33 folds

between us.

Only the wind moves
faster.

I reach 7 folds.

I keep going

but the paper is too thick to fold further.
I need to leave this place:

The impossible space between
us,

the moon
and then some,

everything I write,

every choppy line,
every sentence fragment,
every half-finished poem
carved into the page,

every dead gay boy.

For you,
Adam,
for this,

I will risk everything.

•

                              I will risk everything.

                                           For this,
                                              Adam.
                                            For you.

                                  Every dead gay boy

                            carved into the page,
                         every half-finished poem,
                          every sentence fragment,
                              every choppy line.

                                     Everything I write

                                       and then some.
                                            The moon,

                                                 us,
                       the impossible space between.

                                    I need to leave this place
                    but the paper is too thick to fold further.

                                         I keep going.

                                       I reach 7 folds.

                                             Faster.
                                   Only the wind moves

                                         between us.

                                          33 folds.
                                 I fold and keep folding.

If you fold a piece of paper in half enough times
you'll reach the moon.

You leave
until
I don't believe you.

As you tell me this,
you hold me

by the same unwinding thread
we are strung along,

forgetting
we are alone.

We pretend
we become nothing.

Wake me when
I reach the moon.

Wake me when
the world unfolds beneath us,

when
it's
all
finally
over.

CONTRAPUNTAL TO NOTHING

it isn't when you jump
no
you die
the night before   with your last memory
a mantra   unfolding in real time
from a lost god   with the wind
     that once kept you awake
     with a door that opens to
another door   nothing
     this nothing
you pass through   that you are

you ask me
if i still love you
and for once i'm honest:
you are everything adam
you are everything   that moves
that burns   through me
     that was almost music
a hall of broken mirrors   a prayer for
the abscence of   light
the unspeakable
space   between each breath
yes adam yes
this is true   this is who you are
even when i forget
even when i pretend   you are a boy
you are a god   and you are nothing

(ABOUT SPACE)

i.

Each year, **gravity pulls** the moon 1.6 inches further from Earth. **You** won't live **to** see it, but the moon will eventually abandon us, leading to changes in **the** earth's tilt, falling **sea** levels, extreme temperatures, and mass extinction. **Everything is quiet** in space; gravity is a silent killer.

ii.

According to the law of conservation, matter can neither be created nor destroyed. Applying this law to consciousness gives us the following conjecture: **you** don't **disappear; you** don't **die. You** just **change form.**

iii.

No one knows **how** the universe began; no one knows how it will end, if it will end. Compelling theories exist, but scientists **do**n't have a definite answer.

iv.

Outer space is an ever-expanding, empty void, with only 0.0000000000000000000042 percent of it containing any actual matter.

Let go, **if** you can, of your worries – most of this is **nothing**.

Everything that is, everything that **ever** was, everything we once knew, begins and **ends** in darkness. Can you fathom that?

GRAVITY

at a maximum height of 746 feet, the Golden Gate Bridge is the most popular suicide destination in the world. accounting for gravity, it took six seconds to reach the water from the place you jumped.

6.

you kiss me
and you're gone. six months
after you left
to study physics,

i sift for clues
in research articles
published long before
your death.

5.

as if endlessly reaching
for god's fingertip,
it is impossible
for two atoms to touch.

since they share the same charge,
electrons on the outside
of atoms repel each
other. technically speaking,

the closest one gets

to touching something
is hovering just
above it.

4.

i could write
a metaphor for water,
compare the ocean
to god –

say
in the seconds
before impact, you found
yourself in its image,

your arm
outstretched
& shivering in the
kinetic midnight air.

to do so, however,
would imply
you never actually
touched it.

3.

the night you jumped,
it was cold enough
for the sea

to almost

freeze. there,
the current slows with
each moment, as if
each molecule

is an interlude
in your own death.
as your palms hover
just above

the water, i imagine
them, still warm,
cradling a birthday candle
between your lips.

a soft breeze.
your breath
melting in the air
forever.

2.

of course,
time never
really freezes. only,
the larger an object grows,

the longer each second lasts.
in this space —

i have time to ask you why.
i have time to find your mother.
i have time to write this poem.
and another. and another.

what comes from smoke
is more smoke.

in six seconds,
i have spent years
waiting
for your return.

1.

despite centuries of research,
physicists are woefully unable
to explain gravity.
although undocumented,

it's believed that gravity
has an equal and opposite force
somewhere in the universe.
in this way,

we are never truly apart.
somewhere, a place exists
where the air
does not heat

and the sea does
not thaw and
you are still

frozen in time

as you were once,
wings endlessly spread.
truthfully, Adam,
my pen

is the only force
keeping gravity from
killing you
a second time.

i don't fight
for extra seconds;
i just write the clock
differently.

each day, i close my eyes and
count down until
again
you are right here.

sometimes
i swear
if i reached out at night
i could graze your fingertips

with my own,
wet with longing, as if
you were a god, as if
each finger

were a passing wave
on your skin. but just
as i remember

atoms can't touch,

my hand slips
and again
there is nothing
but moist air

and darkness.
even though
i am always
disappointed,

i still hope
they find you
by the morning,
or at least –

0.

iv.
NOTHING

The night you died, 2,592,000 stars exploded somewhere in the universe.

There is nothing in the end.

Nothing.

Nothing

except

an empty apartment, an impossible dream, and the night I keep reliving.

It's winter. Your heater is broken. But it's ok, we're ok. Until the sun comes out, we'll keep warm in each other's arms.

ESSAY ON LEAVING
after Hunter S. Thompson

Boxing glove on the dashboard
        Cigarette butts in the glovebox
My uber driver
        – who works part-time as an MMA fighter –
punches the gas pedal
        and drives 60 miles per hour
through downtown Portland.

I am leaving your apartment for the first time.
        A breeze
lingers behind me.

We pass the abandoned warehouse.
        The college campus.
Angela's billboard.
        For six months,
she has been searching for a donor,
        her left kidney hollow and lifeless
beneath her skin.

        The billboard is in the densest,
richest part of town.
        She will run out of breath
before she runs out of money.

Two years later,
        the playground where we first met
will be demolished,
        the park paved over and replaced
by a strip mall.
        I will leave Portland,
my last memory of home

a boy swinging back and forth,
        legs lost in constant motion,
flying
        with the northern mockingbirds.

        Swing sets
were always my place for inspiration.

They still are.

•

Kinetic, undressed skin.
My hand
gliding across your naked spine.
In the right light, all shivering boys look electric.

If I time it just right,
the space will align perfectly,
and I'll reach between each atom
in your chest,
blending our flesh together.

I will undo everything
— I swear —
until you sing yourself
to life.

I will meet many men after you,
each one hollow and lifeless
beneath their undressed flesh,
and repeat
the same desperate ritual
with them all.

•

I will leave them how I left you –
        nothing becoming nothing.

•

Leave like you can outlive this body.
        Leave until you outlive leaving.
Leave this memory behind.

At night, I leave the door unlocked,
        the last *if.*

·

                                That winter,
                              I returned home
                    and the billboard had disappeared,
           replaced by an advertisement for a strip mall
                                                two miles west.

                        I bet we had the same blood type.
                              I bet the driver
                    could still swing a nasty right hook.
                              I bet the birds
                                          are still there,
                              their eggs have hatched,
                              and all my unborn friends
                                  have finally begun
                                                to fly.

                              I bet you could have lived.

                                    •

   Maybe it meant something.

  Maybe not, in the long run,
        but no explanation,
no mix of words or music or memories
       can touch that sense of
knowing
that you were there and alive
      in that corner of time and the world.

    Whatever it meant.

NOTHING AND THEN NOTHING

i had a dream, once,
that we were arguing
about something,
and in the distance
there was a black hole
slowly sucking us in.

our bodies began
to stretch.
the thread
began to
unwind.

two spaghetti strands
coming closer
and
closer
to
oblivion.

and we died
while touching
each other,
and nothing
else mattered,
nothing
except
who we were
in that moment
as we reached
our end,
nothing
except

the end,
everything
we once knew
withering
away
to
dust,
and i have
never
loved you
more
than in
that
dream
we shared
together.

this is how it goes.
just like this.

nothing else
matters
now.

i love you, Adam.

alive —
and always.

ELEGY IN FOUR ACTS

In the end:

An entrance,
a placeless place,
an endless
lost language,
Portland's winding streets,
time's crooked fingers,
an unsuspecting sugar pine
and the ivy
lazily encircling it,
an empty apartment
and your cold, darkened silhouette
dissolving
in an unfathomable
sea of nothing.

•

We hold each other
and sink
into the groaning
memory foam.
The sheets wrinkle
like tiny waves.

I tell you
about the bottomless pit,
the empty void,
the music.
I tell you
I was raped.
You grab my hand,
and the world spins
the tiniest bit
faster. Endless
and aching motion.

One of us
will outlive
this night.

The other
will relive it,
over and over.

Do you see it, Adam?

Do you see
the thread
unwinding?

We have passed
the unspoken door,
taken ourselves
beyond where bodies can go.

Everything
is
exactly
how we
imagined.

•

A black hole,
a bullet hole,
and a third eye.

Gravity
holds all of us
together.

NOTES

(four facts about space) borrows ideas from various research articles and Sabrina Benaim's *gravity speaks*.

Aside from its biblical connotations, the phrase "god becoming / flesh. Boy becoming / blasphemy" is inspired by the phrase "Child becoming smoke." from Michael Lee's *The Study of Words and Heaven*.

The phrase "the rituals and doctrines we uncover and keep close" from *Getting High in Your Car* is borrowed from a poem by Ejay Watson.

From the Internet // From My Notes App is a found poem from notes I wrote on the night of my rape. The background borrows phrases from a reddit comment section.

No is an erasure from a reddit post I made.

Gravity adopts ideas from Michael Lee's *The Law of Halves as Applied to the Blade*.

The line "Swing sets / were always my place for inspiration. They still are." from *Essay On Leaving* is pulled from my friend's college admission essay.

Essay On Leaving borrows a quote from Hunter S. Thompson and ideas from Michael Lee's *Decay and More Decay*.

The phrase "endless / lost language" from *Elegy in Four Acts* is adapted from "my eternal lost language" on page 29 of my chapbook, *EJAY*.

ACKNOWLEDGEMENTS

Firstoff, thank you thank you fucking THANK YOU to the following organizations for previously publishing some of these poems, often in completely different forms:

The Big Windows Review – "ELEGY (i.)" (called "Unfold" in this book)

Imperial Death Cult – "About My Teeth" (called "Before You Loved Me, I Hated My Teeth" in this book)

Kalopsia Literary Journal – "Gravity", "The Law Of Conservation [in Four Acts]"

Northern Otter Press – "Essay on Leaving"

Opia Mag – "They Hid in a Closet"

ok, wow. I have so much to say. I don't know where to start. I'm just so grateful that I've written this book, that it's finally out, that I'm still alive to see it. None of this came easy.

To start: thank you, Reader (I'm owning the corniness here), for reading my work and trusting me with your mind, heart, and time. I hope, in some way, these poems connected with you.

Thank you, Ami, Patrick, and Logan, for always supporting me and my work. Ami, your poetry is so fucking cool. Like, seriously, I'm obsessed with it. Keep writing.

Thank you, Ejay, for being my spiritual companion, my essence friend, my rock, my miraculous, for keeping me sane and embracing my absolute batshit craziness too. Thank you, Ejay, for Everything. I love you.

Thank you, Dad, for reading my shitty first drafts, and for reminding me that *art is temporary*. Thank you for always having faith in me, as an artist and as a young adult. I know you raised me the best you could. I forgive you. Please give yourself the same luxury.

Thank you, Mom, for every honeygrow / blaze pizza / sweetgreen dinner date, for loving me unconditionally as I grow and change and continue to drive you crazy, for being the sweetest person I know, and for giving me way too much money over these past two years. You inspire me, as a writer and as a person.

Thank you, Hayden, for being the least (most) crazy person in our family, for putting up with my asshole-ness, for always being eccentric and clever and kind.

Thank you, Christine, for sending me *packages*, for your irresistible snarkiness, for your groundedness, for being a caring, compassionate, hilarious sister-in-law.

Thank you, Elisa, for being my first muse. You're the reason I started writing poetry. I care about you, deeply.

Thank you, Xiu, for your excellent cover design for *EJAY*. and the *Jupiter Review*, for your gorgeous artwork, for your patience as I second-guess myself and ask way too much from you. You're one of the most talented artists I know; certainly the most talented I've had the privilege of working with.

Thank you, Luka.

To so many of my teachers, coaches, and mentors: McDonough, Jacque, Olivia, Don, Jennifer, Bethany, Tooka, Aaron, Alex, Terese, thank you, all of you, for your endless patience.

Thank you, Julia, Ty, Xylo, Jolly Wrapper, Hadiyah, Naomi, Jordan, all my friends from the Portland poetry scene, for giving me my second home. I wouldn't be a writer without y'all.

To my ex(es): I still love you. Not in that way. Nice try. I love you because without you, this book would be absolute dogshit.

To my friends: Erin, Shaun, Sasha, Gianna, Viv, Hector, Sitanya Face, Shell, Theo, Emma, Cara, Jo, Ella, Tasha, Amy, I love you. So much. I'm sorry for being a shitty texter.

Maycee: Remember when I sent you all those poems in 2020? Well, they're here now. Thank you. You were one of my biggest advocates, and that means more than you know.

Lovie: I know it's cliché to say this about a pet, but clichés are cliché for a reason: when I was younger, when I felt so isolated, it seemed you were the only person who truly *got* me. You loved me for me and nothing else. How can I say anything more? You sweet, sweet petunia. Enjoy your rest.

To my 16-year-old self: you did it. What the fuck. When you started writing sophomore year, you'd put all your poems in the same document – first titled *Jupiter* and later *Gravity* – and pretend they were all part of the same book. You never thought you'd actually *publish* it. And you did it. Breathe.

To my current self: This book comes out on 11/11/2022 – your 20th birthday. Please slow down. I love you.

I've spent six months on these four pages. Every time I think it's finished someone new pops into my head. If I don't stop, this will go on forever. No matter who you are, if we've connected in some way – virtually, in person, or otherwise – I haven't forgotten you. Genuinely.

Time is an illusion; nothing ever ends. Please blame my publishing deadline.

ARI LOHR is a queer poet and English Education major at Boston University. Xe is a *Brave New Voices* semifinalist, *Slamlandia* finalist, *Portland Poetry Slam* champion, and a 2021 *Best of the Net* nominee. Focusing on the mystical intersections between power, sexuality, and identity, Ari's poetry appears in the *Northern Otter Press, Opia Lit, Incandescent Review*, and more. They are the author of *EJAY.*, a confessional love letter / poetry chapbook, and *Gravity*, their debut full-length with *Gutslut Press*. They are also the managing editor of the *Bitter Fruit Review* and the editor-in-chief of the *Jupiter Review*. Xe believes truth is malleable, professionalism is violence, and arrogance is sexy. Ari can be found at arilohr.com or @arilohr on twitter and instagram.

ALSO BY GUTSLUT PRESS:

Agender Daydreams – Thokozani Mbwana

baby ur a PIECEofWORK – Ami

Melanin: Black – Dre Hill

Clownstar – Marshall Woodward

Drifting Bottles – Arden Hunter

The Bonemilk Collective: Volumes 1, 2, & 3

@GutslutPress
gutslutpress.com